Pool House Sloths

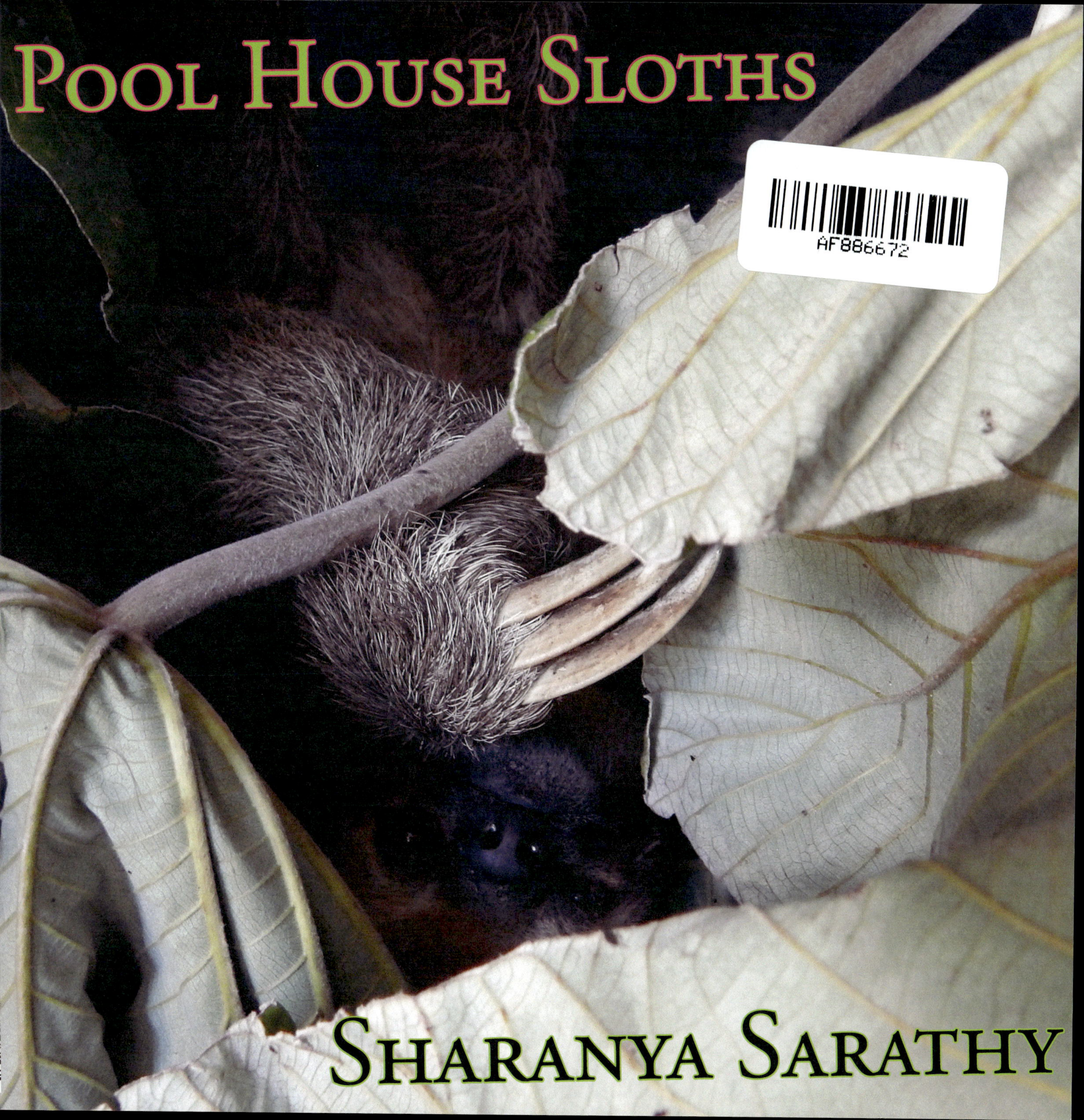

Sharanya Sarathy

For The Wild Sloths In All Of Us
And All The Lucas's
In The Amazonian Rainforest

with immense thanks to

Monique Pool
Green Heritage Fund Suriname
and their team of volunteers

First Edition Published 2013

Photos:
Daya Photography
Monique Pool - page 6 and Sloth Rescue image on page 31

My name is Lucas. I'm a "three-toed" sloth in Suriname, which is a country in South America. I live in Monique Pool's house and this is a book about my friends.

In the rainforest, sloths spend almost all their time hanging upside down in trees and babies like me are normally with their moms. We only come down off the tree once a week to pee and poo. Because we're so slow, algae grows on our fur and that's how we hide from predators. My mom used to tell me to watch out for harpy eagles, jaguars, pythons, and people. Those are our only predators.

Adult male three-toed sloths have a patch on their back, but females don't. When I grow up, I will also have a patch.

Here are some differences between me and my two-toed cousins:

THREE-TOED SLOTHS	TWO-TOED SLOTHS
Species: Bradypus	Species: Choloepus
3 fingers	2 fingers
active day/night (diurnal)	active night (nocturnal)
7 to 9 neck bones	6 to 7 neck bones
have tail	no tail
grayish-brown	light brown
eat only fresh rainforest leaves and flowers	eat leaves, flowers, fruits, and veggies

All my friends at the Pool House are rescued sloths. One of my friends is Naina. Naina loves to just hang around, mostly by the window. We all have a favorite spot to sleep. No matter where we are in the house, we go there to sleep.

My other friend is Anne, who is quieter than Naina. Anne loves to sleep a lot and sometimes even behind a chair.

Because three-toed sloths eat leaves and flowers from the rainforest and almost nothing else, we get special cecropia leaves at the Pool House. Sometimes, we all try to eat the same leaf because the freshest ones taste the best.

Once in a while, we also get special treats like cecropia flowers and fruits, hibiscus flowers and goat's milk. We each have our own favorite treats.

Eating makes us sleepy and so we all have a nap after our meals. Naina likes to sleep behind the tub. Anne sleeps by the lamp or cupboard. I sleep behind the tub or on the window.

Naina likes to try new things. She always reaches for the nearest thing she can grab onto. I am not so bold, but I love to watch Naina and her antics.

We are weighed everyday to make sure we are growing well. I love sitting in the weighing tub. Of the three of us, Naina is the biggest as she drinks milk everyday. I don't like the taste of milk and grow much less rapidly.

Danitsjia is another sloth who lives in the Pool House. She was brought here after escaping from the people who took her from the forest. Though she loves to hang around with us, she always goes back to her favorite spot -- a window in a bathroom.

Harvey was rescued just before he was about to be sold. He only stayed for a short time and once he was eating and drinking well, he was released into his rainforest home.

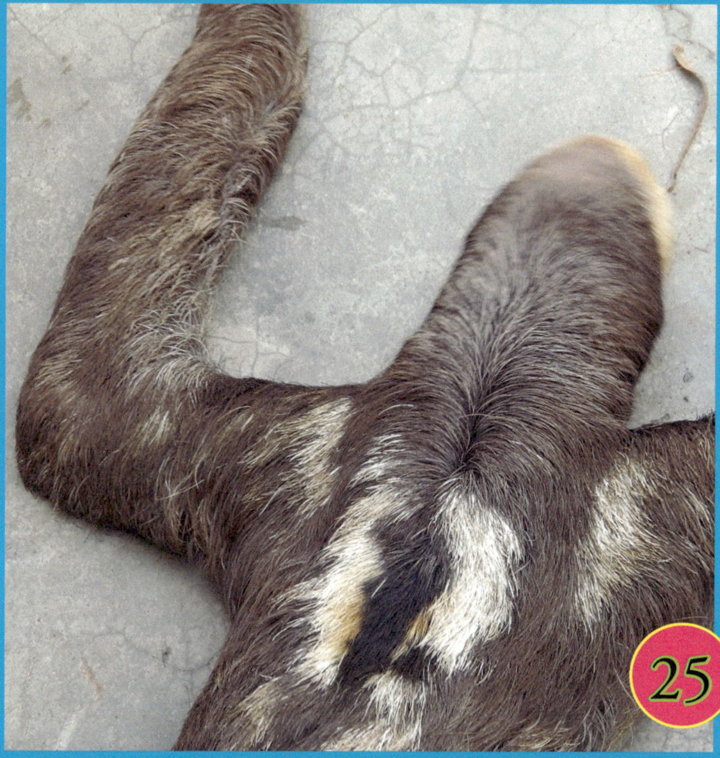

Soekie and Unasur are two other friends who were with us briefly. They were released into the rainforest soon after they were rescued.

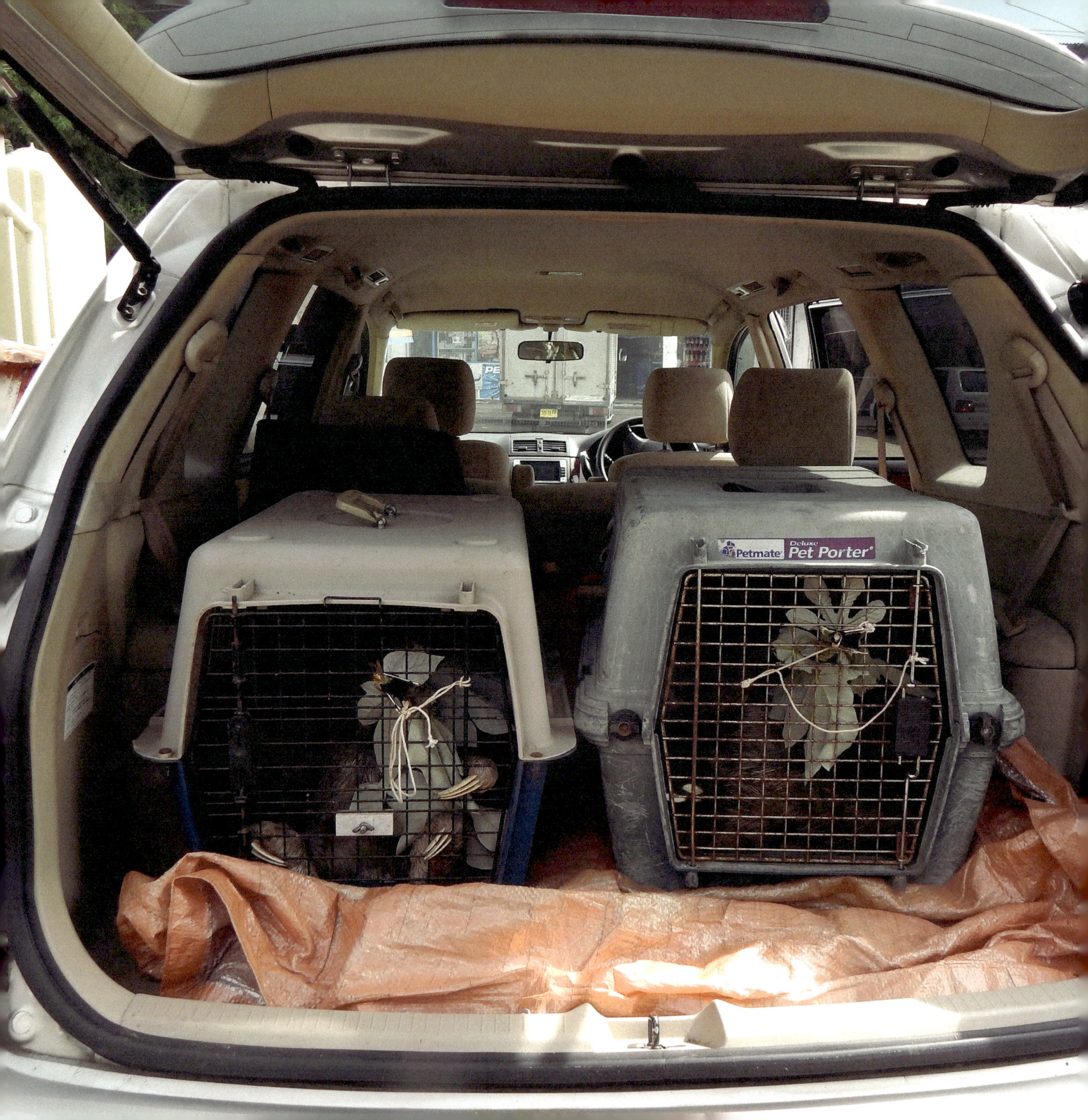
Petmate
Deluxe
Pet Porter

When our friends are ready to be released, Monique takes them deep into the rainforest and lets them go. They happily climb up to the top of the rainforest canopy, where they are completely back at home.

Sometimes our rainforest home is cut down by people for lumber, farms, or cattle pasture. If Monique finds out in time, she rescues us, makes sure we are healthy, and then releases us to a different part of the rainforest. In 2012, she rescued 200 sloths from one place within a few days! She's our hero!!

More pictures of us hanging around ...

... at the Pool House.

Though we love recovering at the Pool House, we are eager to join our friends in the rainforest. I hope to join my friends there someday.

This is me with Sharanya.

This is me with Monique and her sloth socks.

The author contributes proceeds from sales of this book to

Green Heritage Fund Suriname.

Made in the USA
Las Vegas, NV
07 November 2026

57650030R00026